ANIMALS OF THE GRASSLANDS

BURROWING MEERKATS

THERESA EMMINIZER

PowerKiDS press

New York

Published in 2021 by The Rosen Publishing Group, Inc.
29 East 21st Street, New York, NY 10010

Portions of this work were originally authored by Therese Harasymiw and published as *Meerkats*. All new material in this edition authored by Theresa Emminizer.

Editor: Theresa Emminizer
Book Design: Michael Flynn

Photo Credits: Cover, p. 1 pittawut/Shutterstock.com; (series background) Chantal de Bruijne/Shutterstock.com; p. 5 Andrew Lilley/Shutterstock.com; p. 7 Grant Tiffen/Shutterstock.com; p. 9 Alta Oosthuizen/Shutterstock.com; p. 11 Pereslavtseva Katerina/Shutterstock.com; p. 13 mylifeiscamp/Shutterstock.com; p. 15 Roy Michael/Shutterstock.com; p. 17 Albie Venter/Shutterstock.com; p. 19 DragoNika/Shutterstock.com; p. 21 mrinal Pal/Shutterstock.com; p. 22 Susan Schmitz/Shutterstock.com.

Library of Congress Cataloging-in-Publication Data

Names: Emminizer, Theresa, author.
Title: Burrowing meerkats / Theresa Emminizer.
Description: New York : PowerKids Press, [2021] | Series: Animals of the grassland | Includes index.
Identifiers: LCCN 2019049454 | ISBN 9781725316904 (paperback) | ISBN 9781725316928 (library binding) | ISBN 9781725316911 (6 pack)
Subjects: LCSH: Meerkat–Juvenile literature. | Meerkat–Behavior–Juvenile literature.
Classification: LCC QL737.C235 E46 2021 | DDC 599.74/2–dc23
LC record available at https://lccn.loc.gov/2019049454

Manufactured in the United States of America

CPSIA Compliance Information: Batch #CSPK20 For Further Information contact Rosen Publishing, New York, New York at 1-800-237-9932

CONTENTS

Marvelous Meerkats 4
A Meerkat's Home 6
What Do They Eat? 8
Building the Burrow 10
Little Pups 12
Meerkat Mobs 14
Staying Safe 16
Sound the Alarm! 18
Special Adaptations 20
Fancy Features 22
Glossary 23
Index 24
Websites 24

Marvelous Meerkats

Meerkats are cute, furry little animals with pointed noses, round eyes, and long tails. When standing on their back legs, they're about 10 to 14 inches (25 to 35 cm) tall. They weigh 1.4 to 2.1 pounds (620 to 969 g).

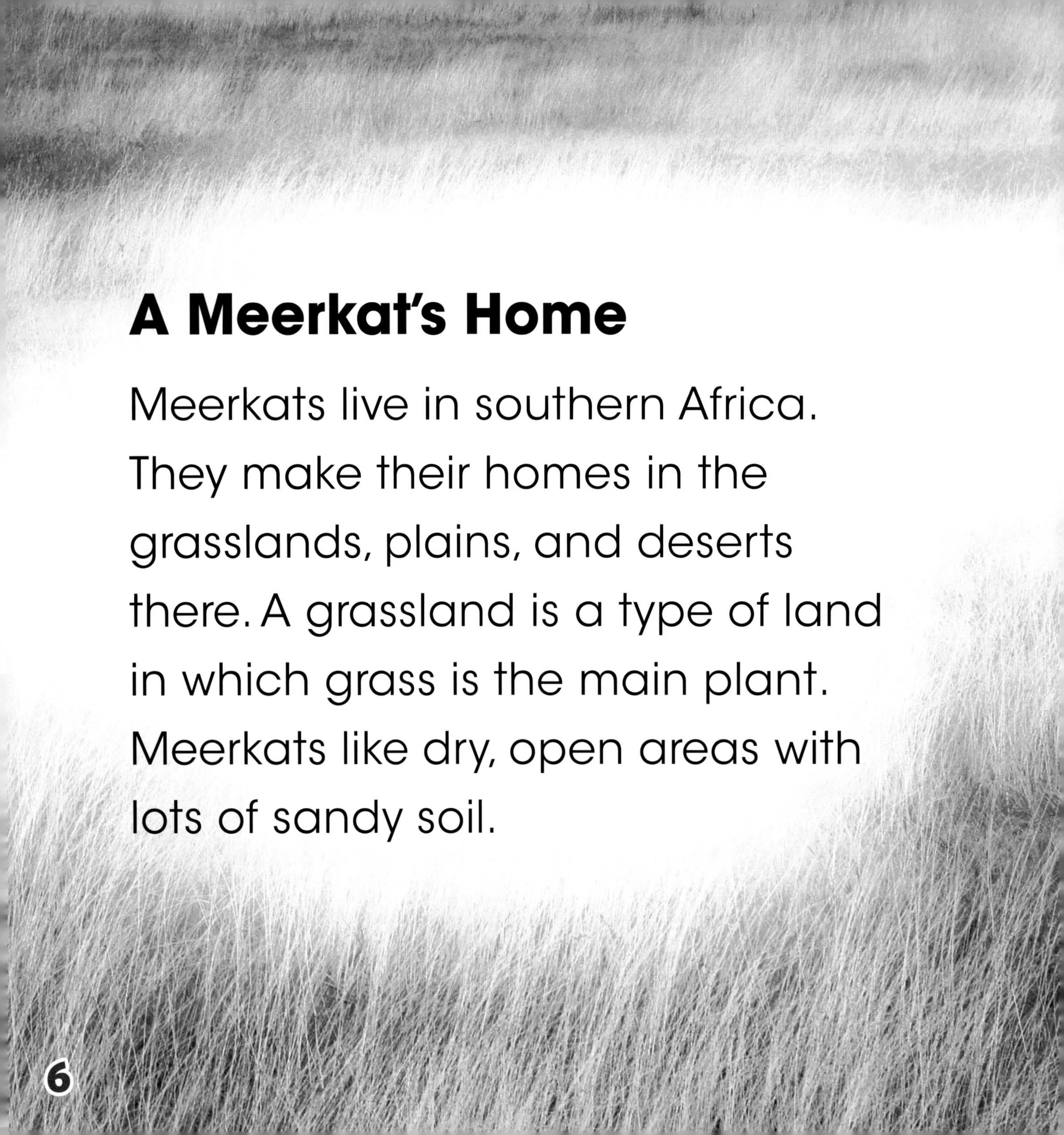

A Meerkat's Home

Meerkats live in southern Africa. They make their homes in the grasslands, plains, and deserts there. A grassland is a type of land in which grass is the main plant. Meerkats like dry, open areas with lots of sandy soil.

What Do They Eat?

Meerkats mostly eat bugs, such as beetles, spiders, and snails. They also eat bigger animals, such as birds, lizards, and snakes. Their strong sense of smell helps them hunt. They use their sharp claws to dig for **prey**.

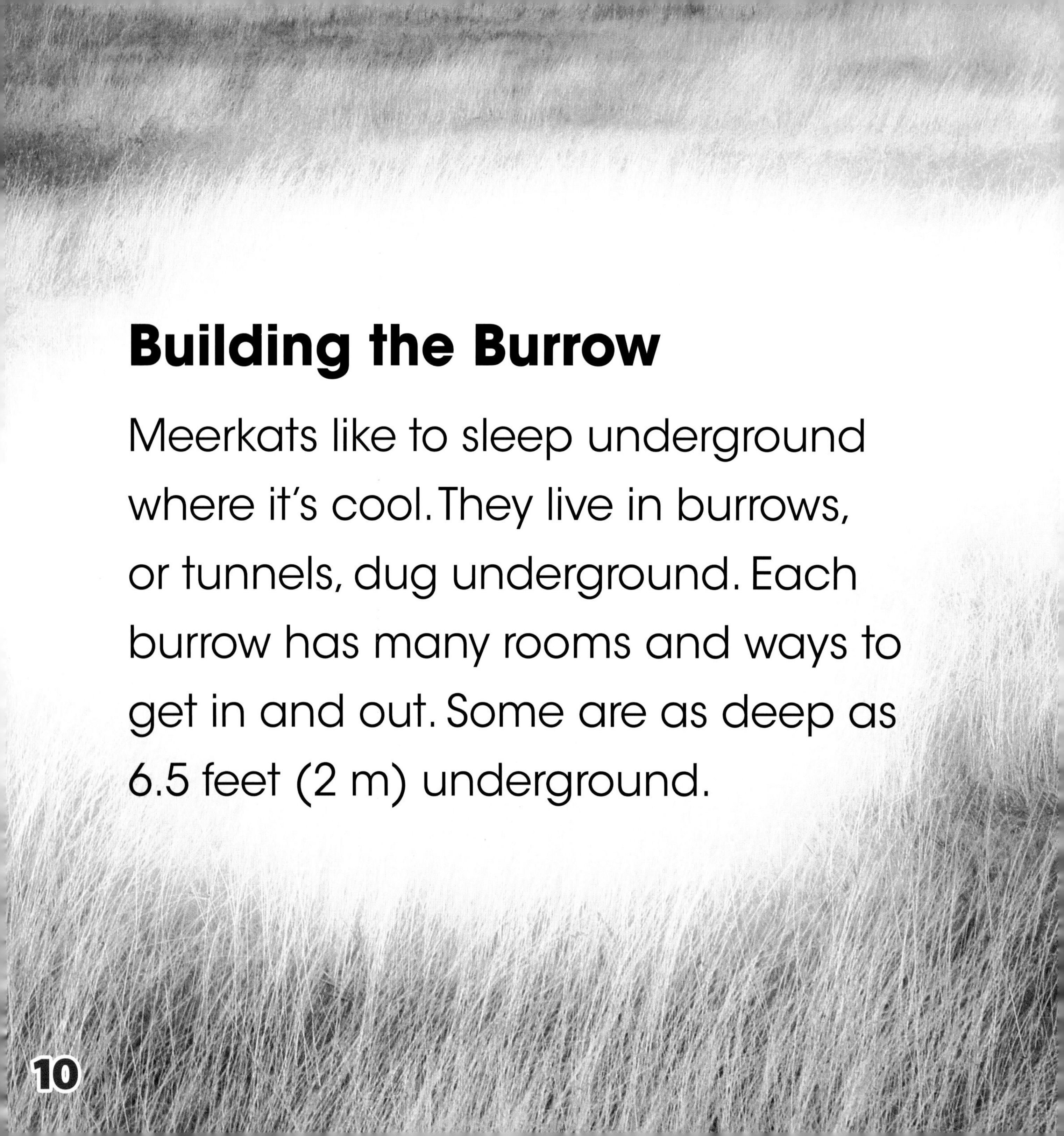

Building the Burrow

Meerkats like to sleep underground where it's cool. They live in burrows, or tunnels, dug underground. Each burrow has many rooms and ways to get in and out. Some are as deep as 6.5 feet (2 m) underground.

Little Pups

Baby meerkats are called pups. They are born in litters, or groups of babies. There are usually two to four pups in a litter. Pups leave the burrow when they're a few weeks old. Wild meerkats live for up to 14 years.

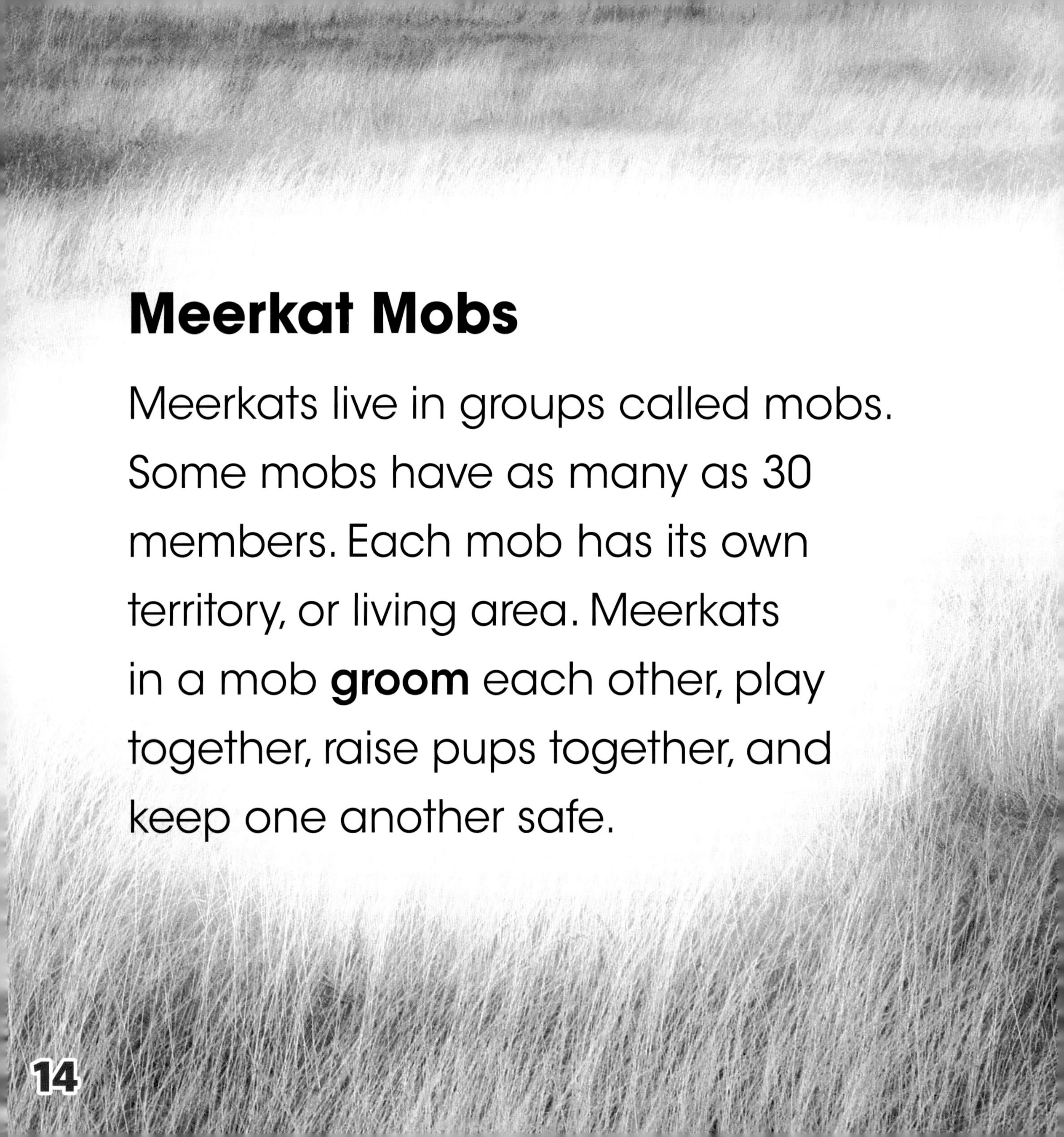

Meerkat Mobs

Meerkats live in groups called mobs. Some mobs have as many as 30 members. Each mob has its own territory, or living area. Meerkats in a mob **groom** each other, play together, raise pups together, and keep one another safe.

Staying Safe

Hawks, eagles, and other animals like to eat meerkats. Meerkats take turns watching over the mob as they look for food. The **guard** meerkat finds a high place, stands on its back legs, and keeps an eye out for **predators**.

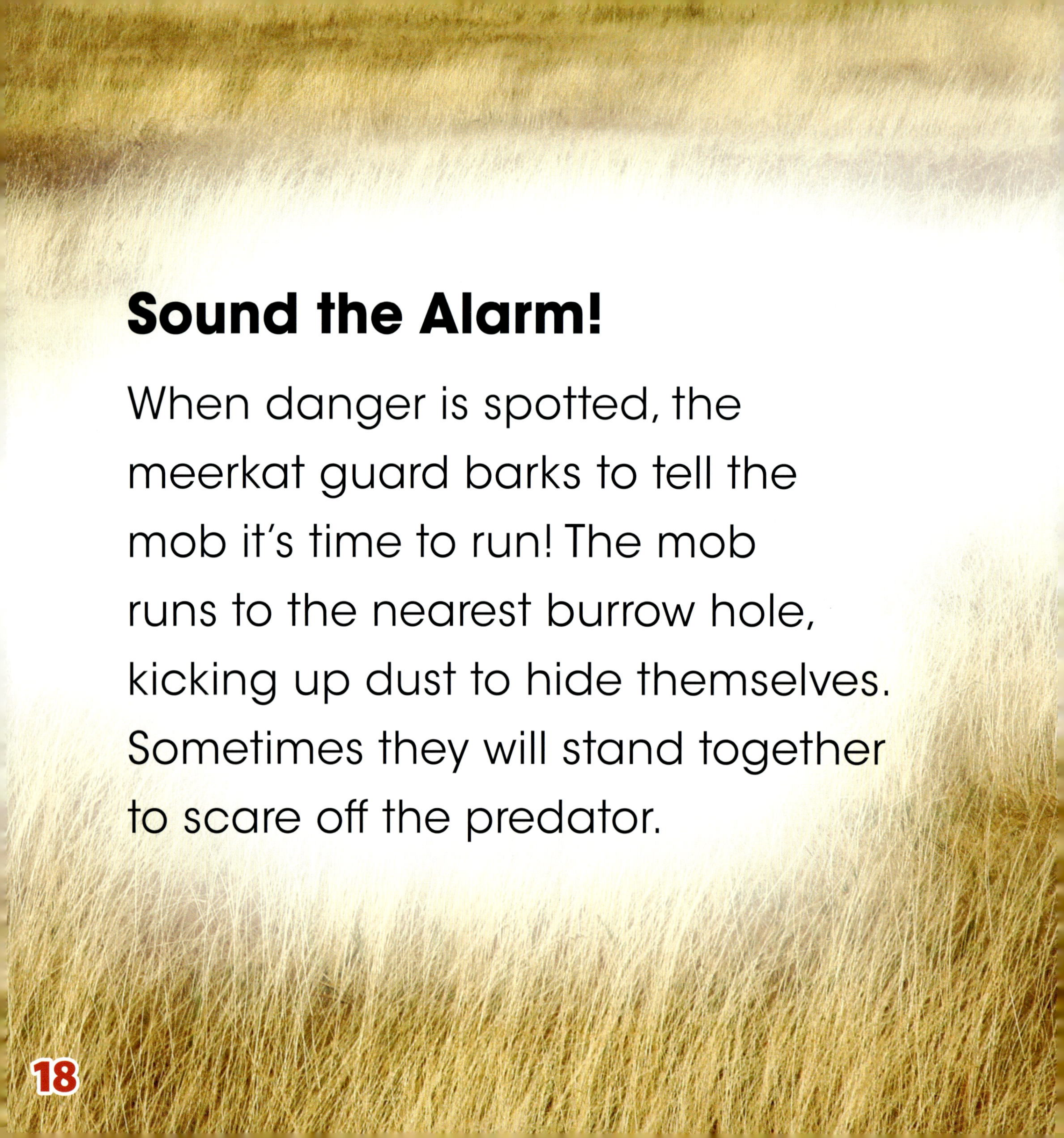

Sound the Alarm!

When danger is spotted, the meerkat guard barks to tell the mob it's time to run! The mob runs to the nearest burrow hole, kicking up dust to hide themselves. Sometimes they will stand together to scare off the predator.

Special Adaptations

Meerkats have adapted, or changed to live better in their grassland home. They can live without drinking water since they eat roots and other foods that hold **moisture**. They stay safe and cool inside underground burrows. Most importantly, they work together to stay alive!

Fancy Features

Height

10 to14 inches
(25 to 35 cm) long

Eyes

special **membrane** to
keep out sand

Weight

1.4 to 2.1 pounds
(620 to 969 g)

Sharp Claws

0.8 inches (2 cm) long

Tail

7 to 10 inches
(18 to 25 cm) long

Feet

five toes on each paw

GLOSSARY

groom: To clean the body and make it neat.

guard: One that watches over others.

membrane: A soft, thin layer of living matter.

moisture: Water in something.

predator: An animal that kills other animals for food.

prey: An animal hunted by other animals for food.

INDEX

B
burrows, 10, 12, 18, 20

F
food, 8, 16, 20

G
guard, 16, 18

H
homes, 6, 20

L
legs, 4, 16
litters, 12

M
moisture, 20

N
noses, 4

P
predators, 16, 18
prey, 18
pups, 12, 14

T
tails, 4
territory, 14

WEBSITES

Due to the changing nature of Internet links, PowerKids Press has developed an online list of websites related to the subject of this book. This site is updated regularly. Please use this link to access the list: www.powerkidslinks.com/aotg/meerkats